ONE SMALL SQUARE®

Cactus Desert

by Donald M. Silver

illustrated by Patricia J. Wynne

LEARNING TRIANGLE PRESS

Connecting kids, parents, and teachers through learning

An imprint of McGraw-Hill

New York San Francisco Washington, D.C. Auckland Bogotá
Caracas Lisbon London Madrid Mexico City Milan
Montreal New Delhi San Juan Singapore
Sydney Tokyo Toronto

Every plant and animal pictured in this book can be found with its name on pages 40–43. If you come to a word you don't know or can't pronounce, look for it on pages 44–47. The small diagram of a square on some pages shows the distance above the ground for that section of the book.

For my friends
Terry and Jimmy Corcoran

We wish to thank Dr. Craig S. Ivanyi of the Arizona-Sonora Desert Museum and Dr. Thomas M. Donnelly for their detailed comments about desert life. We are grateful to Karen Malkus, Maceo Mitchell, and Thomas L. Cathey for their always appreciated efforts.

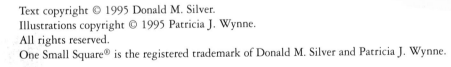

Library of Congress Cataloging Number 97-074148
ISBN 978-0-07-057934-7
MHID 0-07-057934-2
17 LWI 22

Whether you are in a desert or at home, always obey safety rules! Neither the publisher nor the author shall be liable for any damage that may be caused or any injury sustained as a result of doing any of the activities in this book.

Introduction

Whew! It's hot: 115°F (46°C), and the temperature is still climbing. The dry earth bakes under the blazing sun. There's not a cloud in the sky to shield the land from the scorching heat. And not a creature stirs. Does this sound like a place you'd like to explore? You bet!

Welcome to cactus country. Home of rattlesnakes and roadrunners, yuccas and chuckwallas, tarantulas, tortoises, and toads. Welcome to a desert without camels or endless sand dunes. A desert that is as beautiful as it is dangerous.

In this desert you will discover animals that never take a drink of water and plants that double as "apartment houses." You may meet up with a velvet ant that isn't an ant or a beetle that stands on its head. You will witness how heat and dryness can be every bit as deadly as poisonous fangs and sharp stingers.

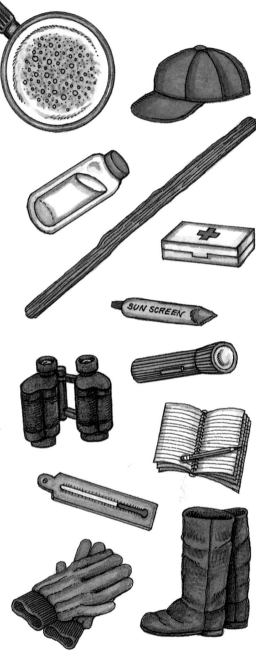

Exploring a desert can be a great adventure, full of surprises. But you need to be prepared in case one of those surprises turns out to be a scorpion, a Gila monster, a tarantula, or a rattlesnake. Before heading out into a real desert, explore one small square of a cactus desert in this book. You will find out when and how to safely explore a desert so that neither you nor the desert is harmed. You will discover the way a desert works, as well as how hard it is for plants and animals to stay alive there.

As you follow along, you will come across activities you can do in your home town that will help unlock some of the secrets of desert life. You will also find activities to do in a desert when you get the chance to visit one. For most of the activities, you will need only the simple equipment shown on this page.

What's Living in a Cactus Desert?

Funguses

Plants

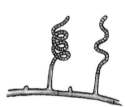

Monera

Animals

Protists

A desert is no place to take chances. Along with your notebook, magnifying glass, and other equipment, be sure to take a hat, gloves, hiking boots, a first-aid kit, sunscreen, a flashlight, and enough drinking water.

1. NEVER explore a desert alone.
2. ALWAYS explore with an adult.
3. Tell another ADULT where you are going and when you expect to return.
4. ALWAYS carry WATER for drinking. Find out the phone number of the local POISON CONTROL CENTER.
5. NEVER explore during the heat of day or late at night. Explore in the EARLY MORNING or EARLY EVENING. But always watch out for rattlesnakes.
6. Wear a HAT and BOOTS for protection. You may also want to wear sunglasses and use sunscreen.
7. NEVER reach into a hole. There may be an animal inside that bites, stings, or stabs.
8. NEVER turn over or pick up a stone. You might frighten a creature hiding under it.
9. LOOK for animals or spines before you sit or kneel.
10. NEVER explore when it is raining or when it might rain. Deserts can FLOOD rapidly.
11. NEVER touch desert animals or plants.
12. TAKE NOTHING from a desert. LEAVE NOTHING behind.

One Small Square of a Cactus Desert

Who knows? One day you really may find yourself in a desert. After all, millions of people live just a few hours' drive from a desert. And millions more spend their vacations near a desert.

You don't have to explore an entire desert to discover amazing plants and animals. One small square will do. The square shown here is 8 feet (about 2.5 meters) on each side. That may be about the size of a bathroom in your house.

Before setting off for a desert, STOP! Check the SAFETY FIRST list on this page. Then you will be ready to search for a small square the size you want. Take a stick —at least as long as a yardstick or meterstick—with you and use it to draw a line on the desert floor that marks off your square. Watch where you draw the line: You don't want to scare a snake or injure a cactus.

No two deserts are exactly the same. Any one you choose will amaze you even if you don't see all the creatures that live in or visit this small square.

Come spend a day and a night in cactus country. Then watch the magic that only water can work in the aftermath of a desert storm. The desert landscape may be one of the roughest, but desert life is some of the toughest.

Beetle eaters
beware: If a dark-
ling stands on its
head, run the other
way, or you'll get a
dose of the beetle's
foul-smelling spray.

All plant roots take in water. But not
all grow alike. A cactus's roots spread
out close to the surface. A mesquite's
roots dive deep into the soil.

Want to know why NOT to put your hand into a hole in a saguaro cactus? Just look in the circle. The boot-shaped holes are perfect for animals to hide away from other animals and from the sun.

Breakfast brings out insect-eating birds and other insect eaters.

Silverfish

Bedbug

Mosquito

Gnat

Pseudoscorpion

Move over, big guys. There's room enough for all. The little creatures in the circle may be cactus dwellers too.

A saguaro "bleeds" when a hole is made—not blood but sap, which hardens and stops the "bleeding."

The Giant

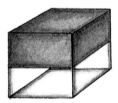

If you've never seen a saguaro cactus, you are in for a shock! It can grow as tall as a tree—up to 50 feet (15 meters). It can weigh as much as an elephant.

One of these cactus giants towers over the small square. It has no leaves, just lots and lots of sharp spines. Instead of branches, it has bent "arms." And it is full of holes.

The "arms" are a clue that the cactus has been around for quite a while. A saguaro doesn't grow arms until it is at least 50 years old. The holes weren't always there, either. Woodpeckers and gilded flickers made them.

Year after year, these birds peck holes for their nests. When nesting time is over, they move out with their young, leaving the holes empty. Empty, that is, until an owl, a flycatcher, a spider, or some other animal moves in. Soon so many creatures are living in the cactus that the saguaro doubles as a desert "apartment house."

When you visit the desert early in the morning, you may catch some "apartment" dwellers out for a snack. Look also for lizards and other guests unafraid to climb the giant.

In May and early June you can see something you'll never forget—a saguaro cactus in bloom.

Your Desert Notebook

There's so much to see in a desert that you'll want to record. So carry a notebook and a pen or pencil with you to draw pictures. If you can't visit a desert yet, write down in your notebook what you find out when you do the activities in this book.

Spaced Out

Use a measuring stick or tape to find out how close a group of trees and bushes grow to each other in your backyard or in a park. When you visit your small square in the desert, measure the distance between plants there. Why do you think desert plants grow farther apart?

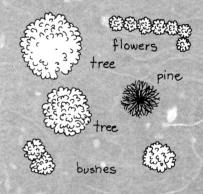

Taking the Heat

Things are heating up in more
ways than one. From out of
nowhere, it seems, a roadrunner
makes a mad dash into the small square and snatches up
a tasty whiptail to eat. A hungry ground squirrel eyes a
circling hawk to avoid becoming a meal itself. And the
hot rays of the midmorning sun deliver a clear warning:
It's going to be another scorcher.

A rattlesnake has had enough. It slithers under a
stone ledge to find relief from the heat. A chuckwalla
crawls into a rock crevice where the sun cannot touch it.
But a collared lizard is still on the hunt for a snack. It

Hawks are predators—they hunt and eat other animals for food.

turns and faces directly into the sun. Stupid move? No! Cool move! Now the sun's rays strike only the top of its head and back, not all along its side.

Soon, though, the collared lizard will sense that it can no longer stay out in the sun. Unless the lizard seeks shade, it will overheat until its body stops working and it dies. Like every other desert animal, the lizard can only take so much heat before it must cool off.

While animals can move out of the sun, plants cannot. They are rooted in place. The secret to how desert plants can take the heat is water. Water keeps plants from wilting. Water keeps plants healthy. With water, carbon

Light colors are cooler. They soak up less sunlight and heat than dark colors.

Fur and feathers protect skin from the hot sun.

Talk about being in a tight spot! A chuckwalla puffs itself up so no animal can unwedge it.

Cactuses can hold on to water in the desert heat. You can't. You lose body water when you sweat and breathe out. So carry water with you and DRINK IT.

dioxide, and energy from the sun, plants can make food. But where's the water? The last rain fell months ago. The earth is like powder. This desert isn't just hot; it is also dry. In fact, all deserts are dry for most of each year.

Stand in your small square and look at the plants. If they are alive, then you have found water in the dry desert. There are tons of water stored as juicy sap inside a saguaro and all the other cactuses. These plants' thick, waxy outer walls stop water from being lost. Sharp spines protect cactuses from many hungry and thirsty animals. And sunlight that reflects off spines never has the chance

There's plenty of sun, plenty of space, but not enough water for more plants.

Leaves are the food factories in most plants. Cactuses have no leaves, yet still make food in their green, water-storing stems.

The Sonoran pronghorn antelope is endangered.

Waxy wall

Juicy insides

Spine

to overheat cactuses.

If you see bunches of dead-looking stalks, they may be live ocotillo or creosote plants. They also trap what little water they need to stay alive.

Is there a mesquite tree full of leaves in your square? Leaves have tiny openings that can lose lots of water. Yet the mesquite can still take the heat. Its roots tap water hidden deep underground, far below other plants' roots.

Whenever you explore the desert, be sure to remember you're an animal, not a plant. Get out of the sun before you feel dizzy or tired and can no longer take the heat.

Cool It

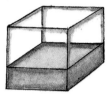

It's mid-afternoon. Not a bird perches on the saguaro's arms. Good thing: the air there is 120°F (almost 50°C). That's hot. No jackrabbit sits atop a rock. Good thing: the air there may be 150°F (about 65°C). That's very hot. No lizard scurries along the ground. The surface is a whopping 170°F (about 77°C) or more! That's way too hot for most animals to stand very long.

Why is the ground hot? There are no clouds in the sky to block some of the sun's rays. The air is very dry. More moisture would soak up some of the rays. So would shady trees—if only they could grow here. Where you live maybe half of the sun's rays reach the ground. In the desert,

Roundworm

Red velvet mite

Bacteria

Green alga

Poisonous centipedes and snakes borrow burrows dug by other animals. That's two good reasons to keep your hands out.

Everything in the circle lives in the twists and turns of the saguaro's shallow roots. There they are safe from the sun, but not from each other.

14

almost all do. And when the ground warms, it gives off heat to the air above it. That makes hot even hotter.

Doesn't 80°–90°F (24°–32°C) sound cool and inviting? Many small square animals are way ahead of you. They are born or hatch knowing they must dig their homes underground. No matter how hot the surface gets, the deadly heat doesn't move far down through the soil. So kangaroo rats, pocket mice, and many other desert diggers snooze the day away in cool quarters as the world above them bakes.

And don't forget those holes in the saguaro. All that cactus sap keeps down the heat. No wonder the "apartment house" is full. Its rooms are 20°F (11°C) cooler than the air outside. They are truly air-conditioned.

The tarantula on the left is resting. The one on the right can't move. A tarantula hawk wasp stung it and laid an egg on it. When the egg hatches, the tarantula will become wasp food.

Wasp egg.

The kangaroo rat plugs the entrance to its burrow. Cool air stays in. So does body water the animal loses during breathing.

Picture This

In your notebook draw pictures of the plants and animals in your small square. Use binoculars to see birds atop a tall cactus. Then look in a field guide to deserts—a book with names and pictures that will help you identify desert life. You may have such a field guide at home. If not, try a local library or bookstore.

Some field guides have pictures showing the size and shape of holes that lead to the homes of different desert animals. Remember: the animal inside may not be the one that built it.

kangaroo rat

wood rat

tarantula

wolf spider

Surprise!

Did the heat finally get to the collared lizard? No, it is fine. As the sun sinks, the lizard turns its side toward the sun and tries to soak up as many rays as possible. Birds make all the heat they need. So do jackrabbits, bobcats, and other mammals—including you. But lizards make very little body heat. The heat they need must come from the sun. When the sun sets, the air cools, and so does the lizard. Before the lizard cools down too much, it turns in for the night.

Baby kit foxes wait in their underground den for their mother to return with food. When they are older, she will allow them to explore their surroundings for the first time.

The collared lizard knows when. The woodpecker does too. And so does the ground squirrel. Every small square creature senses when it's time to come out of the shade and start hunting for food again. For some, that time is late afternoon. Others remain hidden until the cool of the evening.

Surely such a hot desert can't cool much at night. Surprise! As sunlight slowly fades, the scorching heat of the day disappears with it.

Without the sun beating down, the ground loses heat to the air much as a hot pan does when you turn off the burner under it. Unless there are clouds or enough moisture in the air to trap some of the heat, it escapes into space. Most of the time desert air has neither clouds nor moisture. That's why much of the day's heat escapes. Within hours the desert can go from burning hot to chilly to downright cold at times.

Before you return to explore your small square early in the evening while it is still light, STOP! Check the SAFETY FIRST list on page 6 again.

Mark down in your notebook the time you return to your square. Draw a picture of how low the sun is in the sky. Check the plants. Have any of the flowers changed? Are there animals in your square that weren't there early in the morning? Look at the small square shown here. How has it changed? Surprise! The small square at night is a much busier place than it was by day.

A banded gecko lays three eggs and hides them in the ground. Her work is done. But inside each egg a baby lizard is just starting to grow.

Desert honey ants feed on termites and sweet liquids such as flower nectar. But they store some of the liquids for times when food is hard to find. You'll never guess where—inside other honey ants called honeypots. The honeypots swell until they look like they will burst. Then they hang from the ceiling of their underground nest. When needed, they are ready to feed the other ants sweet meals.

Cloud Cover

As evening falls, go with an adult into your backyard or to a nearby park. Take along a flashlight and an inexpensive thermometer. In your notebook record the day, the time, and the temperature. Note if the sky is clear or full of clouds. The next day, check the newspaper or television and record how cold it got during the night. Repeat every day for a week. Do you find that when there are clouds in the sky, nights are not as cold?

While you're out, mark down what sounds you hear. Shine your flashlight and look for animals and plants. Draw pictures of any you see. Are any flowers open? When you visit your small desert square, fill your notebook with the same kind of information. Compare your findings there with those from home.

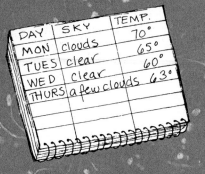

DAY	SKY	TEMP.
		70°
MON	clouds	65°
TUES	clear	60°
WED	clear	63°
THURS	a few clouds	

Going . . . Going . . .

Going . . . going . . . the sun is almost gone. The desert sky still glows with deep, rich colors. Anyone would want to stop and enjoy the sunset. But not for too long. You don't want to miss dinner being served in the square.

Look at the saguaro flowers. The ones that were open earlier in the day are all closed. Those that just opened are serving sweet liquid nectar to bats and moths. Not only is nectar energy-rich, but it is also mostly water. How generous of the cactus to give up precious water to thirsty animals.

Generous? Hardly. As the animals sip their meals, they brush up against pollen grains on male flower parts. Thousands and thousands of grains stick to the animals. When the bats and moths fly to other saguaro flowers for more nectar, they carry the pollen with them. Some grains rub off onto female flower parts. Then the flower can start becoming a fruit full of seeds. Without the help of the nectar feeders, the saguaro could not make seeds. Without seeds, there would be no new saguaros.

Each saguaro flower stays open less than one day. But that's long enough for bats and moths to get their fill of nectar after sundown. And for butterflies, bees, and birds to feast when the sun is up. Enjoy the beautiful flowers while you can. Within hours they will be going . . . going . . . gone.

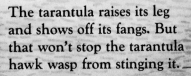

The tarantula raises its leg and shows off its fangs. But that won't stop the tarantula hawk wasp from stinging it.

Looks like a perfect fit—a
long-nosed bat's head inside a
saguaro flower. The bat finds
nectar and winds up with a
head covered with pollen.

Yucca moth   Yucca
              flower
        Pollen

Caterpillar

        Seeds

No yucca moths, no yucca seeds. Why?
Only yucca moths carry pollen from one
yucca flower to another. No yucca seeds, no
yucca moths. Why? The moths' eggs hatch
into caterpillars that eat only yucca seeds.
You can't have one without the other.

In the cool evening, tiny openings in cactus stems let in carbon dioxide gas from the air. At the same time, they let out oxygen gas made earlier along with food.

Finders, Keepers

Here they come: from under stones and inside cracks, from holes and beneath plants. Hungry, thirsty plant eaters are on the move.

The small square has plenty to offer them. To a mule deer, a mouthful of yucca leaves. To a porcupine, palo-verde bark. A wood rat heads straight for a prickly pear cactus. Other animals may fear the sharp spines, but the wood rat uses them as a ladder to reach juicy plant parts. In seconds, water stored as sap inside the cactus winds up inside the wood rat.

Don't expect to see a kangaroo rat bite into anything

Were any munchers and nibblers in your square before you left for the night?

The pocket mouse is another seed eater that rarely, if ever, takes a drink.

juicy, or ever take a drink of water. This little leaper just stuffs its cheek pouches with dry seeds and hops back to store them in its burrow.

How can a kangaroo rat stay alive without water? It can't. But it makes most of the water it needs by breaking apart the sugars in the dry seeds.

Your body makes water from sugar, too. But not nearly as much as you need. Every day you lose lots of water when you sweat, go to the bathroom, and breathe out. A kangaroo rat never sweats. Its body wastes are nearly dry. And during the day it plugs its burrow shut. The seeds stored inside soak up the water the kangaroo rat loses as it breathes. When the kangaroo rat eats the seeds, it gets back the lost water. Water is life: find it, keep it.

What a ruckus! Two kangaroo rats leap into the air in a fight over seeds. For the battle to end, one must hop away.

Hunger brings out beetles, crickets, termites, ants, grasshoppers, and walking sticks in search of food.

21

Stay away from nature's pointed poisoners: a rattlesnake's front fangs, a night snake's rear fangs (in the circle) and a scorpion's stinger.

—Fang

Stinger

Poison

If you've never seen a centipede like this one, you may never want to. It can grow 10 inches (25 cm) long and has a poisonous bite. The good news: It avoids people.

Dangers in the Night

Tarantulas. Night snakes. Need a few more reasons not to explore the desert alone, especially at night? Wolf spiders. Giant centipedes. Scorpions. Still wondering how dangerous a desert can be? Rattlesnakes. Gila monsters. Black widow spiders. What do all of these animals have in common? In a word: POISON.

Poison is scary, but don't let it scare you away from your small square. You probably won't run into all of these poisonous animals in the same place at the same time. Seeing them together here may help you remember what they look like. If you ever do meet up with one, do not disturb it in any way. As long as it doesn't mistake you for a predator, it will leave you alone and go about its business. Animal poisons are usually meant for prey, but some can also kill people.

A snake's pupils open wide at night just like yours do, to let in what little light there is for seeing.

Day

Pupil

Night

Scale

Scales protect skin and keep water in.

The rattlesnake's pits feel the heat (inside the dotted lines) coming from the mouse. They signal to the snake's brain: DINNER!

Rattlesnake poison is very powerful. However, it's of no use unless a rattler can sink its sharp fangs into a rabbit, bird, kangaroo rat, or other snake food. Even in total darkness, a rattlesnake can tell when prey comes within striking distance. The snake may "smell" prey by flicking its forked tongue to pick up odors in the air and on the ground. Or the rattler can pinpoint its next meal with the help of two deep pits on its face. The pits zero in on heat rays coming from birds and mammals. Then, with just one bite, the poison flows. And even if the prey slips away, it's still too late. The poison acts quickly while the snake follows the heat or odor trail to dinner.

The cool of evening soon gives way to the chill of night. Rattlesnakes feel it. So do all the other small square

These spiders are hunters that use poison to kill prey. They may be small, but keep your distance.

23

predators, and the animals they hunt. But not the collared lizard, snug in a burrow fast asleep. Underground the temperature stays a comfortable 75°F (about 24°C), no matter how cold the night gets. And don't forget about those saguaro holes full of woodpeckers and other day-shift birds. All that cactus sap holds in some of the day's heat keeping the "apartments" about 20°F (11°C) warmer than the air outside.

But wait! Something is flying out of the saguaro. It is an elf owl—one of the smallest owls in the world. And he is catching insects on the wing. Not only must he fill his own stomach but he must also return to his "apartment" with enough food to feed his family.

Under cover of darkness, more and more hungry, thirsty predators visit the small square. Although they

Like clockwork, the night-blooming cereus opens its flowers as darkness falls. This plant lives off water stored in its roots.

Hole
Tongue

The kingsnake does fine without poison. Its forked tongue picks up odors that holes in the roof of its mouth "smell."

If it's nighttime, you can be sure some predators are after kangaroo rats.

24

Insects are an elf owl's delight. The owl also hunts scorpions.

Without making a sound, a great horned owl swoops down after a gopher. Lucky for the gopher an escape was so near.

The sound of the coyote reaches far beyond the small square.

may not have poison as a weapon, they come armed with everything from sharp teeth and claws to strong muscles to hearing so keen it can hone in on a mouse stepping on a twig. Each meal caught brings a double reward: food and water. Like your body, other animal bodies are more than two-thirds water.

With both food and water at stake no wonder the desert at night is such a dangerous place. Even poisonous animals are not totally safe, for there are predators that eat them, too. A rattlesnake will shake its rattle to warn a predator to keep away. That often works. Little plant eaters can also get out of tight spots. A kangaroo rat will leap from a predator, turn in midair, and land 15 feet (4.5 meters) away a second later. A pocket gopher can disappear down a hole in a flash. Staying alive in the desert isn't easy, but in nature each animal has a fighting chance.

Signs of Life

What's been happening while you were away from your small square? Check for signs of life: chewed plants, claw marks, animal droppings, and tracks left in the powdery soil. Draw pictures of the tracks in your notebook. Then look in a field guide for a match.

Ringtail

Quail

Cottontail rabbit

Deer

Bobcat

Roadrunner

Dew It

Want to make dew and prove you lose water when you breath out? On a cold day breathe on the inside of a window. The drops that form are a kind of dew. The warm air you breathed out cools so much when it touches the window that it cannot hold all the moisture in it. Some turns into dew. Want to find dew? Search your yard early in the morning for dew on flowers, spider webs, and grass.

Too Cool, Too Hot

Much of yesterday's heat is gone. In the chilly air, some lizards and snakes hardly move. With the first rays of the rising sun, a new day begins.

One by one, creatures of the night slip back into their hiding places. A quick last snack may briefly delay them. On some days, so will a few drops of water.

Yes, water. Sometimes, the desert air cools so much that it can hardly hold even the little moisture in it. When the cooled air touches a rock or cactus, some of the moisture changes from an invisible gas into a liquid. The liquid comes out of the air as drops of dew. Thirsty animals hurry to lap it up. They'd better. Drops falling off of cactus spines disappear quickly into the dry ground full of just-as-thirsty cactus roots. Too bad no dew formed last night in the small square, because today the heat will be blistering.

The sun keeps climbing. At last the collared lizard can start to warm up. It strains to stick its head out of the burrow. Then slowly it crawls onto a rock and turns it side toward the sun. Within half an hour the basking works. The lizard is again able to run after food. As its body colors blend into the rocks and soil, it seems to disappear.

When you return to your square early in the morning, watch the animals closely. In the rush to feed, they will move from sun to shade and back again. Taking time to cool down keeps an animal from overheating. That is, until the sun gets too hot to handle.

Time's just about up for the night shift.
But in the dawn's early light, a bobcat on
the lookout for more food finds it.

It's bath time for the piglike
peccaries. Dust bath, that is—to
get rid of skin pests.

Move away, ringtail.
This is the skunk's
final warning.

If a predator grabs a lizard's
tail, the tail will break. The
lizard can make its getaway
while the surprised predator
is left holding the tail. Later
the tail will regrow.

Butterflies, too,
must warm up in
the morning sun.

27

Following the afternoon rain, the desert feels cool. Clouds and moisture soak up the sun's rays.

Share and share alike? Not when it comes to water. Creosote roots poison other roots that grow too close.

Let's see: In one saguaro fruit there are about 2,000 seeds. The cactus can grow 200 fruits. Take away all the seeds that are eaten, washed away, or stored in burrows, and maybe one seed lands in a place where it can and does grow into a new saguaro.

It pays to advertise. The shape, color, and smell of fruits signal, "Come and get it."

Rain can't save this saguaro.
All that's left is its wood.

Something New

Weeks go by. The creamy-white saguaro flowers are all gone. Nectar is off the menu. Something new is on: juicy ripe cactus fruit. Once again the saguaro needs help of animals—this time to spread the seeds from its fruit. Of course the ants, birds, bats, and kangaroo rats stuffing themselves with seeds don't know that. But some seeds pass through their bodies unharmed and are left in the animals' droppings away from the saguaro.

Something new is also up: clouds. Almost every day they appear and disappear. That is, until late one afternoon, when the clouds keep building higher and higher. Lightning flashes, thunder roars, and strong gusts of wind blow across the desert. The small square darkens. Animals dart for cover. Months of dryness come to a sudden end.

Heavy rains pound the square. The baked ground is so hard that most of the water runs right off it. Just below the surface lurks a thicket of widespread roots, soaking up every precious drop that reaches them.

In less than an hour the storm is over. More than an inch (3 cm) of rain fell—about a tenth of the yearly rainfall. But a few miles away, there was not a drop.

Within days something new is popping out—leaves, on the ocotillo, creosote, and paloverde. Now these plants can make more food. Even the saguaro looks different. You can hardly see the accordionlike folds in the stem. After drinking a ton of water, you'd look plumper too.

Drink Up

Water the small cactus you bought. At the same time water a small houseplant. Place both in a sunny window and let the soils go dry. How long before the houseplant looks like it needs water? How long before its leaves wilt? Rewater at once. And what about the cactus? What happens to plants in your town during a severe dry spell?

Rain Drain

How much rain falls where you live? Make a rain gauge and measure. Ask an adult to cut the bottom off a clear plastic bottle. Take a ruler and mark ½ inch and 1 inch (or 1, 1.5, 2, and 2.5 cm) on the outside of the bottom of the bottle. Be sure to use a permanent marker. Place the rain gauge in an open place and wait until it rains. Did less or more than half an inch (halfway between 1 and 1.5 cm) fall? Does more than an inch (2.5 cm) ever fall? Keep a record in your notebook.

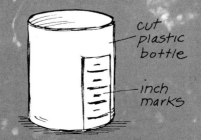

cut plastic bottle

inch marks

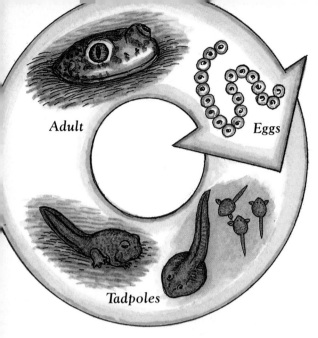

Adult

Eggs

Tadpoles

Spadefoot toad eggs hatch into fishlike tadpoles that swim and breathe through gills. The tadpoles soon grow legs and lungs, changing into adults by a process called metamorphosis, and hunt insects on land. In a few weeks, the toads dig back into the soil until next year.

Seize the Moment

What a wake-up call! To the spadefoot toads in their underground holes, the pounding rain must have sounded like a parade of drummers. Yet that's what it took to stir the toads from their 11 months of deep sleep. The time is right to dig out, feed, find mates, and lay eggs. If they don't, they'll lose the chance for perhaps a year.

The toads seize the moment. They surface as fast as they can. They fill up on insects. And they find what the rain left behind—pools of water.

At once males call to females. The message is clear: Let's mate NOW. And they do, in the water. In just one day the eggs hatch into thousands of tadpoles. For ten or more days the tadpoles grow and turn into toads. Many

After a rain may be the only time that you will see desert millipedes crawling along the ground.

are eaten by predators. The others are in danger because the pool might dry up in the hot sun before the toads are ready to live and breathe on land.

If there's a pool in your small square, the last thing you want to do is step in it. Instead, watch it closely. You may see tadpoles and lots more.

One-celled creatures and algae also have been waiting for rain. So have fairy shrimp eggs laid at the bottom of a pool that vanished just about a year ago. With fresh water the algae come to life, make food, and divide into more algae. The one-celled creatures feed on the algae and on each other. They divide. The eggs hatch into fairy shrimps that grow into adults, mate, and lay more eggs as quickly as the spadefoot toads. Before the pool dries, the shrimps too must seize the moment.

Fairy shrimp

Tadpole shrimp

Green alga cyst

Mosquito larva

Just add water and "presto." Eggs hatch and cysts, hard coverings on one-celled creatures and algae, break open. It's time to feed and breed.

You can get a head start on finding out what kinds of plants and animals live in deserts. Visit a natural history museum and see if it has a desert display. If there is a zoo or a botanical garden near your home, visit that, too. Draw pictures of desert plants and animals in your notebook. Ask how hot the exhibits are kept and how often they are watered. If you own binoculars, take them along. Practice spotting animals large and small.

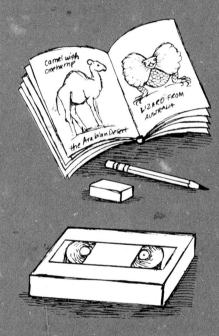

camel with one hump

the Arabian Desert

LIZARD FROM AUSTRALIA

There's always a way to bring a desert into your home—videos. Your local library may have some. Check under nature titles. Try to spot animals and plants in the videos. Practice makes perfect.

31

In a Hurry

The "magic" is still at work. Making the leaves appear was a great opening act. The shrimps and toads were a neat trick. But almost overnight, colorful wildflowers in the desert? That's the magic that only water can work.

For a year or longer wildflower seeds have sat in the soil. Blistering heat did them no harm. The cold of winter nights could not kill them either. Tough seed coverings made sure of that. Rain alone makes the seeds sprout. And not just any rain; only a heavy downpour will do.

Desert wildflowers are in a great hurry. Unlike their seeds, the plants can't take the heat or the cold for very long. A few weeks and they are gone. Once the seeds

Even after a heavy rain, not all desert wildflowers bloom. Each kind waits until it senses the time is right for it to grow, flower, and make seeds. Imagine waiting 50 years to sprout. That's what some wildflowers can do.

Timing is everything. Desert hummingbirds nest when there is lots of nectar to drink and lots of insects to feed their young.

open, roots grow as fast as they can to soak up water. Stems, leaves, and flowers follow soon after. Then the plants must make new seeds before the soil dries out completely, and they die. If they fail, they won't be back next year. That's why desert wildflowers must wait for enough water to fall. Too little and they lose the race to make seeds before the dryness robs them of life.

If you can, explore your small square when the wildflowers are in bloom. Under your magnifying glass you may see beetles, bees, and bugs feasting or carrying pollen. Some of these small animals grow into adults only when flowers are there for food. They, too, are in a hurry to mate and lay eggs before losing the battle against the heat, dryness, or predators. As always, don't you be in a hurry.

Nearly all lizards are predators. But not the chuckwalla. It fattens up on fallen saguaro fruits as well as on flowers and leaves. And the small square has them all.

The velvet ant may be fuzzy like velvet, but it packs a wallop. It's really a stinging wasp.

33

Desert Diorama

Take a shoebox and measure its length and height. Cut a piece of paper for the background wall about ¼ inch (about 6 mm) shorter than the box height and about 4 inches (10 cm) longer than its length. On it you can draw saguaro cactuses and other desert plants. Place the picture in the box and tape each side to the front. The picture will curve.

On separate sheets draw and color desert plants and animals, each with a flap at the bottom. Cut out each picture, bend its flap, and glue or tape it to the bottom of the box. You can also draw and color rocks—or use real ones. Sit a small cactus in your desert diorama to make it look more real. You can create one box for day and one for night.

You don't know what's hiding down in the flowers.

After a cloudburst or two, dry time returns. The hot summer sun speeds up evaporation—water changing from a liquid into a gas that escapes into the air. Within weeks the pools disappear. For a time there is still some water in the soil. It moves up from below, carrying minerals that plants need to their shallow roots.

The small square hurries to prepare for hard times. The Gila monster retires underground to live off its body fat for nine months. A pocket mouse in its burrow falls into a deep sleep that will last all summer. Kangaroo rats and wood rats stock up on as many seeds as they can. The ocotillo and other plants shed their leaves. Fruits fall. Wildflowers die. And many small creatures do, too. Their bodies are slowly broken down into more minerals and other nutrients that return to the soil to be recycled by plants.

Until the next storm, the saguaro must live off its stored water. The longer the giant can stay plump, the better. Still, with no more rain, the cactus stem will start to refold like an accordion. No hurry.

Small Square by Day

Can you match each living thing to its outline?

Collared lizard

Arizona pocket mouse

Western diamondback rattlesnake

Desert wood rat

Prickly pear cactus

White-winged doves

Saguaro cactus

Desert swallowtail butterfly

Anna's hummingbirds

Cactus wren

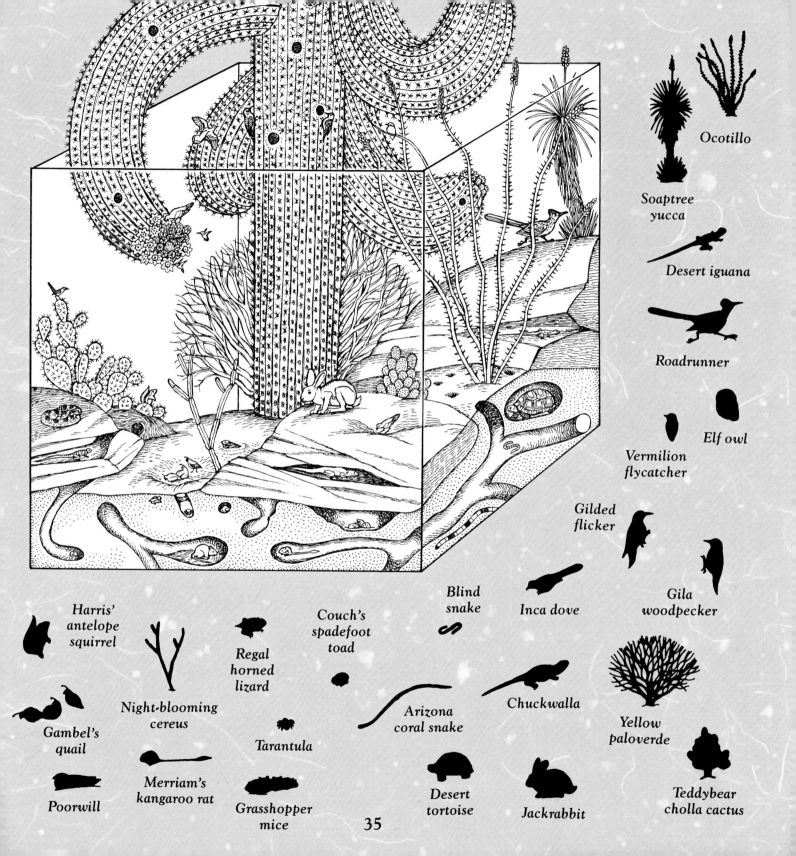

Ocotillo

Soaptree
yucca

Desert iguana

Roadrunner

Elf owl

Vermilion
flycatcher

Gilded
flicker

Gila
woodpecker

Blind
snake

Inca dove

Harris'
antelope
squirrel

Regal
horned
lizard

Couch's
spadefoot
toad

Night-blooming
cereus

Arizona
coral snake

Chuckwalla

Yellow
paloverde

Gambel's
quail

Tarantula

Poorwill

Merriam's
kangaroo rat

Grasshopper
mice

Desert
tortoise

Jackrabbit

Teddybear
cholla cactus

35

Double Trouble

Heat and dryness spell double trouble for anything that tries to live in the small square. Even so, just look at all the different ways desert plants and animals have of overcoming these two dangerous foes. As rough as the desert gets, everything in the square toughs it out.

Well, almost. In most deserts less than 10 inches (25 cm) of rain falls in a year. But there's no guarantee. Some years may be completely dry. Then small cactuses shrivel, and other plants wither. There isn't enough food for plant eaters or their babies. Fewer and fewer plant eaters mean predators also suffer. On a "full tank" of water, a saguaro can last almost two years without rain. After that, it too is in danger of dying. And when it finally does rain, torrents of water can flood the hard-baked soil, harming still more living things.

When you explore your square, keep in mind that the desert is home to the plants and animals you see. Never dig up any plant, no matter how much you like it. Animals depend on desert plants for food, water, shade, and shelter. Don't take seeds, either. They keep the desert alive. Remember double trouble: If you go to the trouble of caring about the desert, the desert won't get into trouble because of you.

The Sonoran Desert, in the southwest United States and northern Mexico, is the only place where saguaro cactuses grow. It has two rainy seasons: summer and winter. These flowers bloom only in the spring.

Saguaro cactus

Long-nosed bat

Long-tongued bat

Sphinx moth

Desert wood rat

Prickly pear cactus

Kingsnake

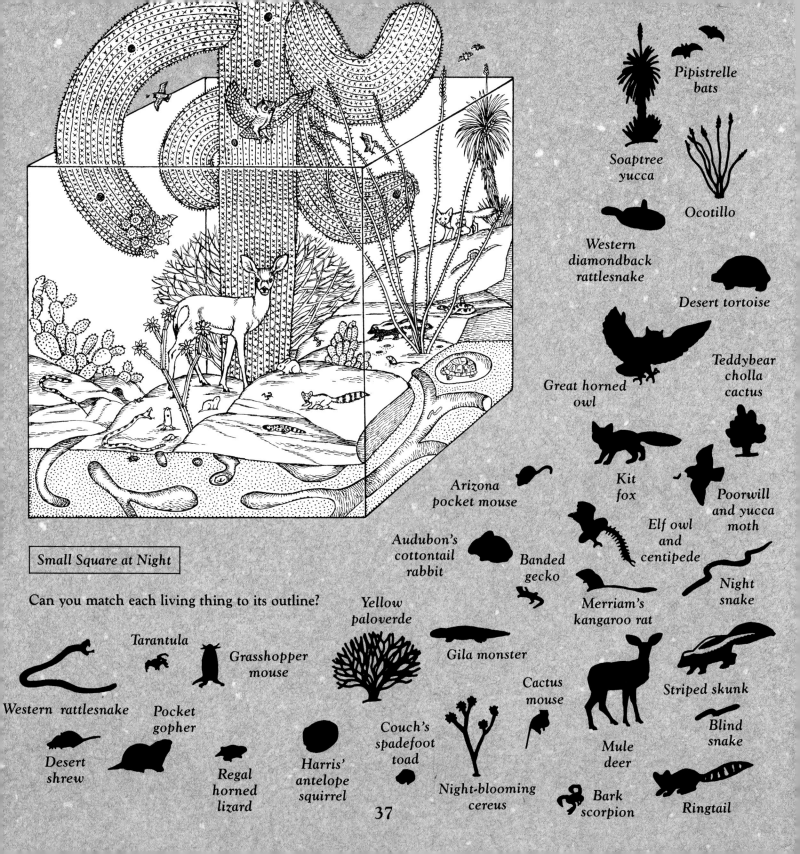

Pipistrelle bats

Soaptree yucca

Ocotillo

Western diamondback rattlesnake

Desert tortoise

Great horned owl

Teddybear cholla cactus

Kit fox

Poorwill and yucca moth

Elf owl and centipede

Night snake

Arizona pocket mouse

Audubon's cottontail rabbit

Banded gecko

Merriam's kangaroo rat

Small Square at Night

Can you match each living thing to its outline?

Tarantula

Grasshopper mouse

Yellow paloverde

Gila monster

Cactus mouse

Mule deer

Striped skunk

Blind snake

Western rattlesnake

Pocket gopher

Desert shrew

Regal horned lizard

Harris' antelope squirrel

Couch's spadefoot toad

Night-blooming cereus

Bark scorpion

Ringtail

37

Deserts of the World

Don't worry about running out of deserts to explore. They cover more than one-fifth of the land on Earth. While all deserts are dry, not all are hot. The Gobi Desert in central Asia, for instance, is cold most of the year.

In the Gobi Desert, in east-central Asia, the first fossils of dinosaur eggs were discovered in 1923. In 1993, fossils were found of a baby meat-eating dinosaur still inside its egg. If it had hatched, the baby might have looked like the one in this illustration.

If it's sand dunes you want, think Sahara Desert in northern Africa. Of course, you may have to do a lot of exploring, since most of this vast desert is flat and rocky. In some parts no rain has fallen in 20 years.

Devil's Hole pupfish

(actual size)

Death Valley, in the western United States, is one of the hottest deserts on Earth. If you ever go there, try to visit Devil's Hole—a small pool where the endangered desert pupfish lives.

There is more water in Antarctica than in any other land. But it is almost all frozen solid as ice. Unless the ice melts, Antarctica will remain the largest frozen desert on this planet.

These desert animals are vertebrates. They have bones in their bodies.

You can tell mammals by their fur and birds by their feathers. Look for dry scales on reptiles, tough scales on fishes, and a thin, scaleless skin on amphibians.

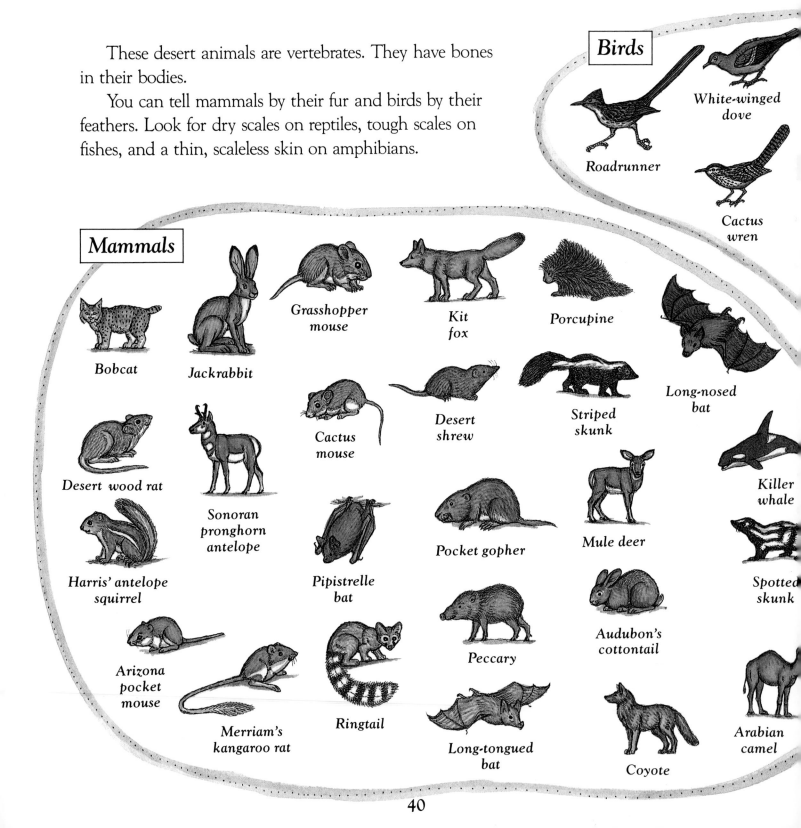

Birds

White-winged dove

Roadrunner

Cactus wren

Mammals

Bobcat

Jackrabbit

Grasshopper mouse

Kit fox

Porcupine

Long-nosed bat

Desert wood rat

Sonoran pronghorn antelope

Cactus mouse

Desert shrew

Striped skunk

Killer whale

Harris' antelope squirrel

Pipistrelle bat

Pocket gopher

Mule deer

Spotted skunk

Arizona pocket mouse

Merriam's kangaroo rat

Ringtail

Peccary

Audubon's cottontail

Long-tongued bat

Coyote

Arabian camel

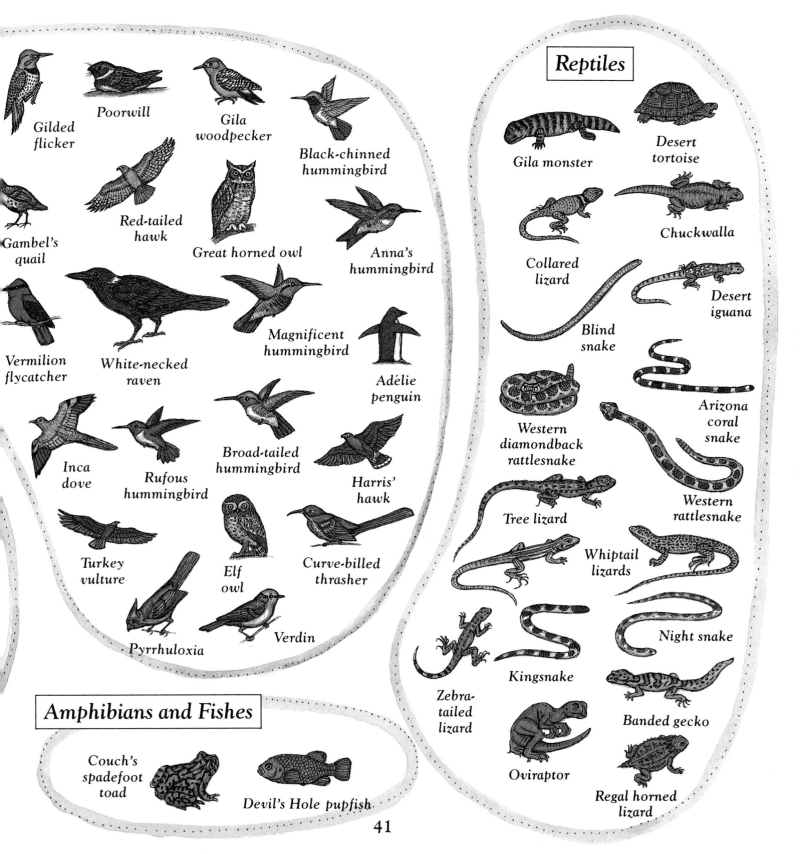

Gilded flicker

Poorwill

Gila woodpecker

Black-chinned hummingbird

Gambel's quail

Red-tailed hawk

Great horned owl

Anna's hummingbird

Vermilion flycatcher

White-necked raven

Magnificent hummingbird

Adélie penguin

Inca dove

Rufous hummingbird

Broad-tailed hummingbird

Harris' hawk

Turkey vulture

Elf owl

Curve-billed thrasher

Pyrrhuloxia

Verdin

Reptiles

Gila monster

Desert tortoise

Collared lizard

Chuckwalla

Blind snake

Desert iguana

Western diamondback rattlesnake

Arizona coral snake

Tree lizard

Western rattlesnake

Whiptail lizards

Zebra-tailed lizard

Kingsnake

Night snake

Oviraptor

Banded gecko

Regal horned lizard

Amphibians and Fishes

Couch's spadefoot toad

Devil's Hole pupfish

41

Insects and other invertebrates have no bones. Most of these desert animals are small. Some are poisonous.

Be sure to look at desert plants and funguses under your magnifying glass. To see one-celled monera and protists, you would need a microscope.

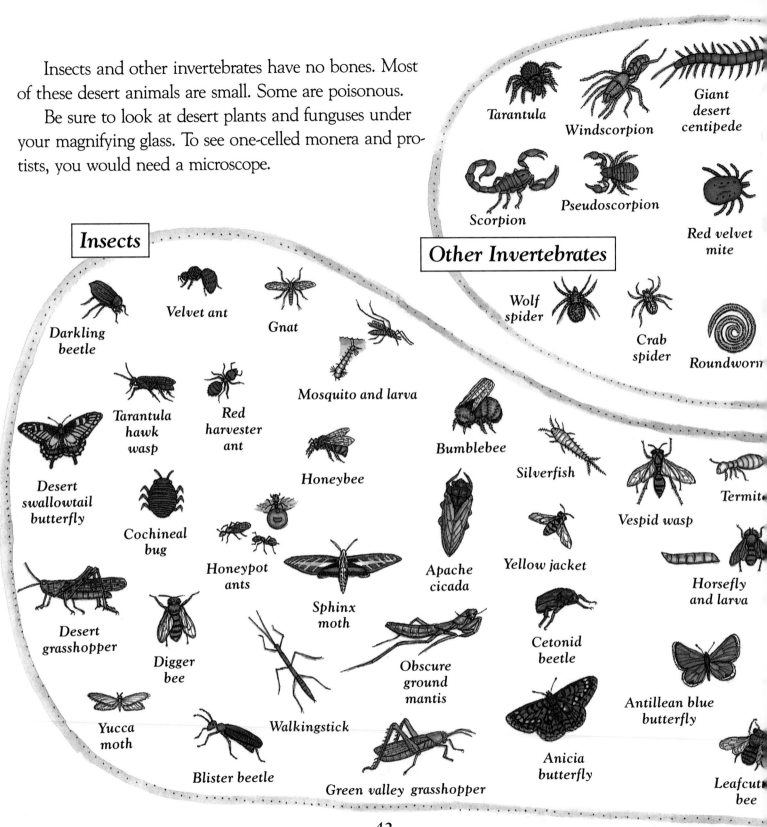

Other Invertebrates

Tarantula

Windscorpion

Giant desert centipede

Scorpion

Pseudoscorpion

Red velvet mite

Wolf spider

Crab spider

Roundworm

Insects

Darkling beetle

Velvet ant

Gnat

Mosquito and larva

Tarantula hawk wasp

Red harvester ant

Honeybee

Bumblebee

Silverfish

Vespid wasp

Termite

Desert swallowtail butterfly

Cochineal bug

Honeypot ants

Sphinx moth

Apache cicada

Yellow jacket

Horsefly and larva

Desert grasshopper

Digger bee

Walkingstick

Obscure ground mantis

Cetonid beetle

Antillean blue butterfly

Yucca moth

Blister beetle

Green valley grasshopper

Anicia butterfly

Leafcutter bee

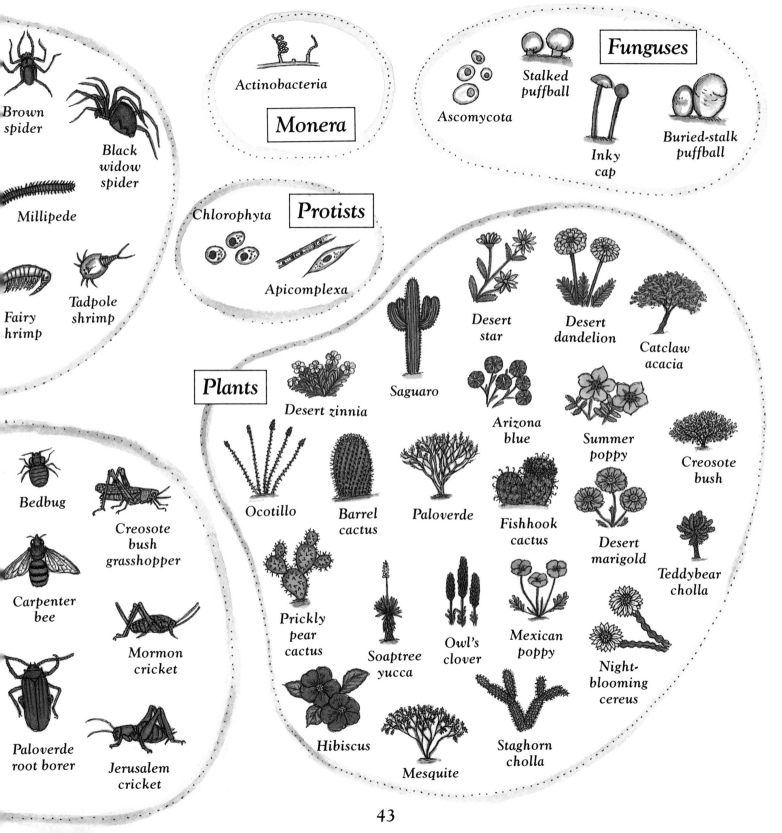

Brown spider

Black widow spider

Millipede

Fairy shrimp

Tadpole shrimp

Monera

Actinobacteria

Funguses

Ascomycota

Stalked puffball

Inky cap

Buried-stalk puffball

Protists

Chlorophyta

Apicomplexa

Bedbug

Creosote bush grasshopper

Carpenter bee

Mormon cricket

Paloverde root borer

Jerusalem cricket

Plants

Desert zinnia

Saguaro

Ocotillo

Barrel cactus

Paloverde

Prickly pear cactus

Soaptree yucca

Owl's clover

Hibiscus

Mesquite

Staghorn cholla

Desert star

Desert dandelion

Catclaw acacia

Arizona blue

Summer poppy

Creosote bush

Fishhook cactus

Desert marigold

Teddybear cholla

Mexican poppy

Night-blooming cereus

43

Index

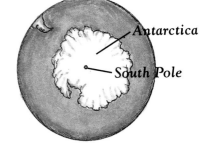

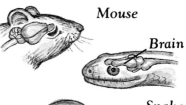

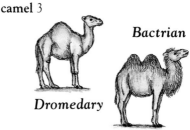
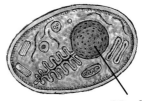

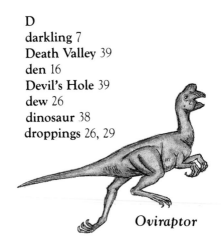

Index

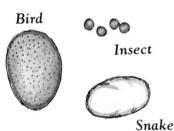
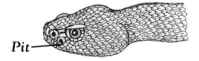
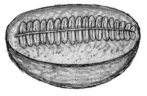
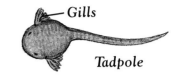
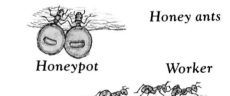

Index

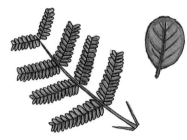
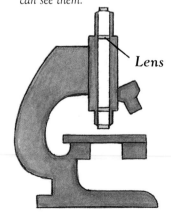
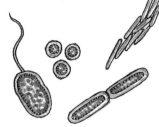

Index

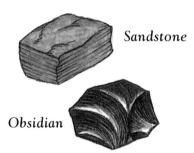

Sandstone

Obsidian

Thundercloud

Sweat gland

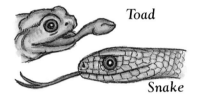

Toad

Snake

Find Out More

Learn even more about the Sonoran desert by visiting the

Arizona-Sonora Desert Museum
2021 North Kinney Road
Tucson, Arizona 85743
(602) 883–1380

Further Reading

Look for the following in a library or bookstore:

Golden Guides, Golden Press, New York, NY

Golden Field Guides, Golden Press, New York, NY

The Audubon Society Beginner Guides, Random House, New York, NY

The Audubon Society Field Guides, Alfred A. Knopf, New York, NY

The Peterson Field Guides, Houghton Mifflin Co., Boston, MA

Reader's Digest North American Wildlife, Reader's Digest, Pleasantville, NY